SOLVING COLEBROOK EQUATION

for

Pipe Design, Sizing and Optimisation Calculations

M Anil Kumar

(MS Chemical, IIT Madras, Industrial Consultant)

Notes

The ability to simplify means
to eliminate the unnecessary,
so that
the necessary can speak

-Hans Hofmann

COPYRIGHT STATEMENT

www.ColebrookEquation.com

CONTENTS

---NOTICE---

By purchasing this book, the purchaser is abide to international copy right rules

1)   to prevent illegal copies of this work entering in to public domain in any form, as computer file or print copy

2)   To use this book if purchased as software copy (PDF) only in a single computer system

PREFACE

Many books exist to elaborate application of Excel for engineers and scientists. Every engineers and scientist use Excel or other worksheet for their professional calculations.

FORTRAN, which was a sequential programming tool, became unpopular for general engineering tasks because it needed sequential intervention and any error in feeding the inputs required that everything is executed again. Excel being a **special calculation** tool provides advantages like special data and inputs (instead of sequential in FORTRAN). Any error in input can be corrected in a click.

A recent survey indicated that majority of Excel users does not like to use VBA or Macros because of several reasons. The **spatial feel** (sense of space that we have in our life) and the comfort is lost. Many have no time to learn macros or rather not consider it worthy. Some fear also exists that macros can erase data and equations or enter values in wrong cells if a minor error occurs in program. I always look for doing things well within worksheet! It is like we always wish to do things *within* than sourcing from outside!

Implicit equations in the form x = g (x) exist everywhere in engineering, especially in transport phenomena (heat, momentum and mass transfer), chemical kinetics, and in certain area of physics and most importantly in fluid dynamics. All scientists and engineers are not great mathematicians and thus finds it difficult to proceed with "complicated" equations where the parameter x appear on the either side of the equation.

We consulted many famous books like Excel Scientific and Engineering Cookbook (by David M. Bourg), Excel for Scientists and Engineers – Numerical Methods (by E Joseph Billo), Excel 2007 for Scientists (by Dr. Gerard M. Verschuuren) etc. to name a few to know how Excel can solve these equations. There is no treatment of iterative solutions *within* worksheet seen in any reference books we could examine as on the publication date..

Engineers need a method to solve thousands of instances within their worksheet with a good degree of simplicity. We worked a great deal on iterations within worksheet, how to introduce a iteration, to seed an initial approximation, to control and to limit number of iterations to a specified times – without using VBA or Macros or *solver* or *goal seek*. We succeeded and this book share you the findings.

This book describes necessary fundamentals for chemical engineers, piping designers, refrigeration experts, ducting engineers, HVAC experts etc. to enable them solve Colebrook equation (and other such implicit equations which converge) in the ease and comfort of a spreadsheet.

ACKNOWLEDGEMENTS

Special thanks to senior colleague of mine, Mr Mehaboob P K, who remained with constant encouragement and motivations. Encouragements extended by friends and by family members and specially patience extended by my daughter Aradhana in the course of my experimentation and writings are always remembered.

Your suggestions to improve this book

-M. Anil Kumar

1   SPREADSHEETS FUNDAMENTALS

Engineers and scientists have to do tremendous amount of computations. All over the world, they use computers for their numeric and scientific computations. Charts and even calculators are out already.

Among the software emerged, FORTRAN66 and its subsequent versions became popular in 1960s as a computational tool. FORTRAN as a computational tool had several drawbacks and inconveniences among scientists and engineers.

- Most of them could not spare time to master FORTRAN.
- FORTRAN needed extensive manual intervention and thus remained inconvenient
- If any error detected in earlier stages of calculation, all subsequent stages required repeated execution.
- FORTRAN did not have appeal of a normal paper based calculations.

Notwithstanding this drawbacks, FORTRAN is still used after half a century of its origin, but manly for advanced computations like that needed in most intensive supercomputing tasks, such as weather and climate modeling, computational fluid dynamics, computational chemistry, economics, and physics.

SPREADSHEETS
Spreadsheets are computational tools that simulate calculations on paper, but highly enhanced user-friendly capabilities. Spreadsheets have multiple cells arranged grid like in columns and rows. Each cell can contain a value (numeric, alphanumeric or alphabet) or equation. Spreadsheets originally became popular for financial calculations mainly because any errors in entry can be corrected by changing value or equation in any cell resulting in *instant recalculation* of an entire worksheet or workbook (combination of worksheets). This features attracted scientists and engineering professionals as well.

Spreadsheet tools arrived in late 1970 itself by the arrival of VISICALC and later tools like LOTUS 123. MICROSOFT EXCEL later emerged as a robust spreadsheet program. Recently so many free replacements for proprietary programs like EXCEL is becoming popular and so is the web based spreadsheets (example, GOOGLE DOCS).

Our further discussions will be with reference to MICROSOFT EXCEL, as a typical, robust spreadsheet program that is currently available in millions of systems across the world.

WORKSHEETS

A spreadsheet file may contain multiple interdependent sheets, that are called worksheets. Each worksheet contains thousands of rows and hundreds of columns and consists thousands of cells arranged in columns and rows like stacked boxes.

CELLS

Each cell is a "box" containing your data. The data remains safe and intact in each box. From another cell, this box can be referred and operated to get a certain mathematical result. For example, if the cell A1 (column A and row 1) contains data 12 and cell A2 contain data 3. You can enter a formula =A1/A2 in any other cell to get result 12/3 which is equal to 4.

To discuss detailed procedures to enter formula is beyond the scope of this book[1]

---

[1]   Refer the book "Making Smart Scientific/Engineering Calculation Worksheets in Excel" by same authors.

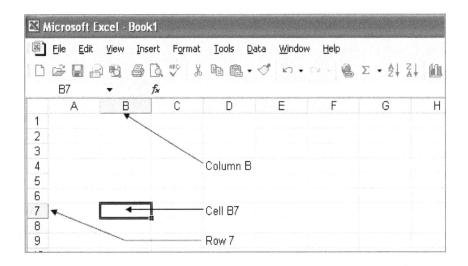

CELL STABILITY

An important point about spreadsheets is that the content in a box (cell) is never modified unless the user manually does so by using an input device like a keyboard. At the same time, a cells value may change if any cell that is referred changes. This gives a stability feel for a spreadsheet.

Once any cell is modified, the whole worksheet is recalculated automatically to update the effect of the change.

ADVANTAGES OF SPREADSHEET PROGRAMS

Spacial Programming: Almost everyone find it easier to perform calculations in spreadsheets than by writing the equivalent sequential program in FORTRAN. This is due to two charecteristics of spreadsheets. Spreadsheets use *spatial relationships* to define *program relationships*. We have highly developed intuitions about spaces, and of dependencies between items. Sequential programming usually requires entering line after line of text, which must be read slowly and carefully to be understood and changed.

Easy Debuggung: Spreadsheets are tremendously forgiving, allowing partial results and functions to work. One or more parts of a program can work correctly, even if other parts are unfinished or broken. This makes writing and debugging programs much easier, and faster. Sequential programming usually needs every program line and character to be correct for a program to run. One error usually stops the whole program and prevents any useful result.

It is often convenient to think of a spreadsheet as a mathematical graph, where the nodes are spreadsheet cells, and the edges are references to other cells specified in formulas. This is often called the dependency graph of the spreadsheet. References between cells can take advantage of spatial concepts such as relative position and absolute position, as well as named locations, to make the spreadsheet formulas easier to understand and manage.

USE OF VBA IN EXCEL

VBA is a "event driven" programming language, VISUAL BASIC which can run codes in host applications like EXCEL. VBA is having immense potential if handled by experts in it. Unfortunately, most of the scientists and engineers are not much familiar with VBA and they do not have patience or time to devot and master the art of VBA. The "Object Model"s in VBA is not transparent to many scientists and engineers. Also VBA is not a spacial programming tool and often need manual intervention to complete tasks.

If a task can be done within worksheet, no one will ever wish to write a VBA program for the same. This is because flexibility of any technique that can be performed within the worksheet is tremendous compared to limitations when done same task in VBA. In this book, we will examine how such a task of solving implicit equation can be done well inside the worksheet itself without any VBA head aches!.

2   NUMERICAL CAPABILITIES OF EXCEL

It is a general impression that Excel is a spreadsheet program that is designed for business calculations– like forecasting, cash flow planning, and budgeting, financial reporting, and so on. One reason for this could be, unlike for the scientific and engineering applications, there exist several source books on business applications of Microsoft Excel

Many research students, engineers and scientists do use Excel for data analysis, statistical analysis, graphing, and other engineering and scientific calculations. There still exist lack of confidence and clear knowledge on capabilities of Excel and similar spreadsheet programs in engineering and science.

It will be a pleasant surprise for any engineer/scientist to know that Excel's specifications and capabilities is far more than needed by any computational requirements in their area. Let us first examine the specific numerical capabilities and limits of Excel.

PRECISION

Number of precision in Excel is 15 digits compared to 10 digit of a calculator. Regardless of the number of digits displayed, Excel stores numbers with up to 15 digits of precision. If a number contains more than 15 significant digits, Excel converts (truncates) the extra digits to zeros (0).

Because of this, the number
12345123451234554321 is only considered as 12345,12345,12345,00000 by Excel.

Similarly,   0.000000111111222223333344444   is   considered   as
$1.11112222233333 \times 10^{-7}$.

This limitation in precision does not mean that the size of number that Excel can handle is 15 digits. With a 15 digit precision, the size of number that can be handled is huge, (for example, $1.23451234512345 \times 10^{26}$)

Some limitation of excel is worthy to mention, though this limitations no way limits any calculations pertaining to any quantity or measurements a scientist may come across.

The number **12.3456789123456**789 is considered numerically as 12.3456789123456 and **-0.123456789123456**789 is only -0.123456789123456

In other words, we can not add 0.1 to 123451234512345.0 and get 123451234512345.1 as answer, as Excel truncates it back to 15 significant digits and so there will not be any change. Also you can not add 12345 to 12345,12345,12345,00000 to get 12345123451234512345 because of same reasons.

DISPLAY AND PRECISION : The display of number of digits and format of display is decided by the number format that you choose for the cell. Changing the display formats does not change the number behind the appearance, the actual numerical value.

You can use number formats[2] to change the appearance of numbers, including dates and times, without changing the numerical value handled by computer. For example, you can display a number such as .08 as 8%. Also, 12345.12345 will be shown only as 12345.1 if you selected number format with 1 decimal place. Still internally the number will be 12345.12345 which are evident if you multiply the cell with 10000, which returns 1234512345.

However, there exist a method (called Precision as Display option) to perform your calculations only as per display format you choose. For example, 12345.12345 will be considered only as 12345.12 if you change the cell to a two decimal format.

---

[2]   Format – cells - number

EXCELS LIMITS IN SIZE OF NUMERALS

This may slightly change depending up on the release version of software but the figures given below are to get an idea of exytreme values that Excel can handle. We need not bother about the exact 15 digit precision values of these limiting values, as such numerals never occur in our day-to-day calculations.

Largest in Magnitudes, Farthest from zero

Largest number that can be typed in is $1.7976, 93134, 86231 \times 10^{308}$

=179769313486231000 !

Largest negative number allowed (ie. farthest from zero) is -1×10^{308}

Smallest in Magnitude, closest to zero

Smallest positive numbers that is allowed in Excel is
1.979×10^{-308}

The largest allowed negative number (ie. closest to zero) is
-2.225×10^{-308}

SCIENTIFIC/ENGINEERING LIMITS

Smallest numbers that finds use in science or engineering should be the measurements in subatomic level. For example, diameter of a nucleus is in range of 10^{-15} m. It can be noted from the above discussions that Excel can handle numbers as small as 10^{-308} which is far less (10^{293} times) than that needed for describing nuclear radius.

Following is a comparison on Excel's capability and smallest numbers needed in science.

Smaller Side

Excel's Lower Limit	$: 10^{-308}$
Planks Constant	$: 6.6262 \times 10^{-34}$ J-sec
Electron Rest Mass	$: 9.10953 \times 10^{-31}$ Kg
Atomic Mass Unit	$: 1.66057 \times 10^{-27}$ Kg
Diameter of Atomic Nucleus	$: 10^{-15}$ m

Higher Side

Speed of light	$: 2.997925 \times 10^{8}$ m/s
Number of cells in our body	$: 10^{14}$
Avogadro Number	$: 6.022045 \times 10^{23}$ /mol
Mass of Earth	$: 5.977 \times 10^{24}$ Kg
Distance to the farthest observed object	$: 10^{26}$ m
Excel's Upper Limit	$: 10^{307}$

Scientists and engineers can relax!

Excel has precision requirements and number handling/processing requirements that far outweighs the most severe requirements in our scientific world. From the above discussions, it is evident that for any of the ordinary calculations needed for calculations involving a real life situation, only a fraction of the numerical capability of Excel is needed. This information is quite encouraging for a scientist/engineer/researcher who decides to use Excel their computational and data management tool.

Above discussions encourage any engineer or scientist to use spreadsheets like MS Excel for their calculations.

3   EXCEL – BASICS

This book assumes that the user has basic knowledge in entering formula and working with cell references. Some essencial points are discussed below. A sample of each essential point is illustrated and the reader is encouraged to do experiments in worksheets to take forward the skill.

REFERENCES
Each cell in Excel has reference or address . Following are examples

Range	Reference
The single cell in column A and row 10	A10
The range of cells in column A and rows 10 through 20	A10:A20
The range of cells in row 15 and columns B through E	B15:E15
All cells in row 5	5:5
All cells in rows 5 through 10	5:10
All cells in column H	H:H
All cells in columns H through J	H:J
The range of cells in columns A through E and rows 10 through 20	A10:E20

ENTERING FORMULA
Formula is entered in each cell directly starting a = sign. Once you enter the formula in a cell, it will be visible in the formula bar right side to *fx* (see below)

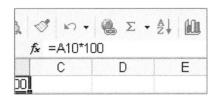

TO REFER CELLS BY RANGE

Cell address can be reffered as reference as given above. For example, refer a value in A10 in a equation like =A10*100 [See the figure below: If A10 has value 12, this equation you entered in any cell will return a value 1200]

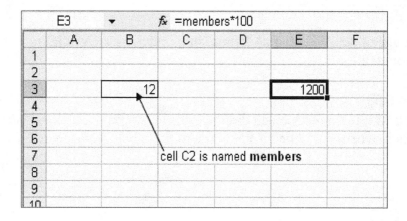

REFER CELLS BY NAMES

It is possible to name[3] each cell by a word of your choice and refering to that word in equation will refer to the value of the named cell. In the following example, the cell C2 is named members and in E3, the equation entered is =members*100

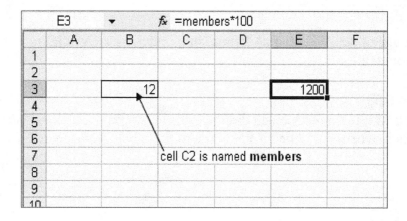

---

[3]   Insert > name > Define >
enter the name and click OK button in the dialogue box.

OPERATORS

Formula works by operators. One should be through in the use of operators to successfully use Excel for computations.

Arithmetic operators: To perform basic mathematical operations such as addition, subtraction, or multiplication; combine numbers; and produce numeric results, use the following arithmetic operators.

Arithmetic operator	Meaning (Example)
+ (plus sign)	Addition 3+3
– (minus sign)	Subtraction 3–1 –1
* (asterisk)	Multiplication 3*3
/ (forward slash)	Division 3/3
% (percent sign)	Percent 20%
^ (caret)	Exponentiation 3^2

The above arithmetic operators are essential in any formula

Comparison operators: You can compare two values with the following operators. When two values are compared by using these operators, the result is a logical value either TRUE or FALSE. There are applications of these operators in IF statements which we will discuss soon.

Comparison operator	Meaning (Example)
= (equal sign)	Equal to

	A1=B1
> (greater than sign)	Greater than A1>B1
< (less than sign)	Less than A1<B1
>= (greater than or equal to sign)	Greater than or equal to A1>=B1
<= (less than or equal to sign)	Less than or equal to A1<=B1
<> (not equal to sign)	Not equal to A1<>B1

FUNCTIONS

Functions which are predefined formulas are used for special calculations like Sine, Log, Cos, Average, etc. There are 329 worksheet functions available in Excel in category like Trigonometric, Logical, Engineering, Financial etc.

Discussing each of the 329 functions and its efficient usage is beyond the scope of this book. For the skill that is going to elaborate in this book, however we need to master certain logical functions.

4   IF LOGICAL FUNCTION

"If" is a powerful logical function that allows many magic in a worksheet and thus need detailed treatment in this book. In our day-to-day we take several decisions based on IF.

For example, IF we get up late, we skip our breakfast to reach office at correct time.

This is represented in following equation.

If (*got up late*, **skip breakfast**, take breakfast)

In this equation, following parts need be clearly understood.

Got up late	*"Condition"*
skip breakfast	*"value, if the condition is true"*
take breakfast	*"value, if the condition is false"*

IF statement in excel is one of the simple and versatile functionality that can be employed in highly efficient calculation algorithms..

"CONDITIONS"

The conditions in the above can be specified by any comparison operators we discussed before like =, >, <, >=, <=, or <>.

Information Functions (next chapter) when used in conjunction with IF, can remove hurdles in using circular reference leading to efficient iterations.

SIMPLE IF EQUATIONS

A simple IF logical function using > as the comparison operator is entered as follows.

$$=IF(A10>0,1,2)$$

If A10 is positive value, then this equation returns 1 and OTHERWISE this equation returns value 2

SIMPLE IF
Instead of values 1 and 2 , we can use cell reference in the above.

$$=\text{IF}(A10>0,\text{A1,A2})$$

NESTED IF
Nested if has multiple IF conditions. For example, if we get up late, skip breakfast and otherwise check what is the beverage (coffee or cola?). If coffee, take bread and otherwise take cake.

If (*got up late*, **skip breakfast**, if (coffee served, take bread, take cake))

Mathematically such statements are called **nested IF** as the underlined set of **if** is nested within the main **if**.

$$=\text{IF}(A10>0,\text{A1,If(B10>0,B1,C1)})$$

If A10 is positive, the cell becomes value in A1. Otherwise (if not) it checks B10 is positive and use B1 if so. Otherwise, the value C1 is used.

MAXIMUM NESTED STATEMENTS IN A EQUATION
Excel can accept maximum of 7 **nested if** in a cell. This does not pose any limitation because each of "Value if true" and "value if false" can be another cell where there is again such many IF statements.

INFORMATION FUNCTIONS AS CONDITIONS WITHIN IF
As already seen, **information functions** bring 'true' or 'false' based on the cell referred. IF can accept this result from information functions like ISERROR and can branch to "value if true" or "value if false".

An illustration is given below

	A	B	C	D	E	F
1	APPLICATION OF ISERR WITH IF					
2						
3	1 Cell D3 is error		=	#DIV/0!		
4						
5	2 Check the cell D3 is error or not			TRUE	=ISERROR(D3)	
6						
7	3 Use IF function to branch accordingly			641.23	=IF(ISERROR(D5),D10,D12)	
8						
9						
10	4 Branch Values			641.23		
11						
12					122.76	

The equation ISERR(D3) returns TRUE. If this is nested inside an IF as shown in the yellow cell entering IF(*ISERROR(D5)*,D10,D12), the branching occurs to D10. So the value 641.23 appears in the equation cell.

Blank Page

5 INFORMATION FUNCTIONS

To detect and avoid errors automatically in a calculation, we need information about cells that return error. This error then can be eliminated once we have information that the one or more cells have returned errors.

Each of the information functions, referred to collectively as the IS functions, checks the type of value and returns TRUE or FALSE depending on the outcome. For example, the ISBLANK function returns the logical value TRUE if value is a reference to an empty cell; otherwise it returns FALSE

Information functions available in Excel are yet another powerful tool that we can employ in sophistication of certain iteration automation. We will be concentrating on information related to errors in a worksheet.

ISERROR FUNCTION

The function is used in the following syntax

$$ISERROR(Value)$$

The 'value' can be a cell name or cell reference.

For example = ISERROR ($A10$). The cell returns TRUE if the cell **A10** is any of the errors listed below.

ISERROR	Any error value (#N/A, #VALUE!, #REF!, #DIV/0!, #NUM!, #NAME?, or #NULL!).

The result of information functions is TRUE or FALSE.

IMPORTANT ERROR TYPES

#DIV/0! Error: This occurs when a number is divided by zero. This error can happen in a iteration if there is a division by x term in $f(x)$ side of iteration, and if in the course of iteration x approaches zero value. Certain automatic way of seeding the iteration with a nonzero value can help to eliminate this, which will be dealt in detail.

.

Error: This is only a display error that occurs when a column is not wide enough (or rarely a negative date or time is used).

Increasing the width of the column, or reducing number of decimals in the number format can correct this error

#REF! Error: When a cell tries to reference a cell that can't be located on the spreadsheet either due to deletion or having been overwritten by pasting by a user or a VBA code, it can result in a #REF! error.
If you get the #REF Error in your iteration sheet, try checking if the cell is using a non-existent cell. Pasting any cell to a cell referred in a equation can cause this error. Deleting a row or column referred by a equation can also cause this error.

Example: The equation =(H5+H6) returns #REF! if you paste any cell on H5 or H6. Same may happen if you delete row 5 or row 6 or column H.

6   CIRCULAR REFERENCE

When a formula refers back to its own dependent cell, either directly or indirectly, it is called a circular reference.

EXAMPLE FOR CIRCULAR REFERENCE

Suppose in cell A1, you enter =B1. Then go to cell B1 and enter =A1. Now cell A1 need value in the cell B1 but the equation in need value from A1.This means both cells A1 and B1 are waiting for results in the other cell causing a deadlock situation in which each cell waits results from the other cell and no meaningful things happen in the above particular example.

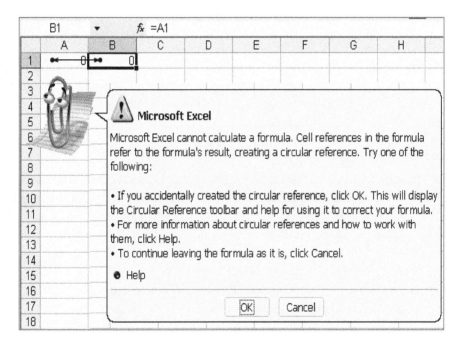

Excel cannot by itself calculate worksheets when it contains a circular reference. A warning message appears as above. Iteration calculation should be activated to perform a worksheet calculation with properly.

The simple circular reference will not result anything because all cells have a default value of zero and simple mutual dependence between each other won't lead to any calculation.

ACTIVATE CIRCULAR REFERENCE

1. On the **Tools** menu, click **Options**, and then click the **Calculation** tab.
2. Select the *Iteration check box.*
3. To set the maximum number of times Excel will recalculate, type the number of iterations in the **Maximum iterations** box. The higher the number of iterations, the more time Excel needs to calculate a worksheet. Maximum iteration is 32,767. In many practical iterations, only one or few hundreds of calculation is sufficient to get reasonable stabilization.
4. To set the maximum change you will accept between calculation results, type the amount in the **Maximum change** box. The smaller the number, the more accurate the result and the more time Excel needs to calculate a worksheet.

NOTE: The maximum change should be less than the increment to continue iteration. When the increment is less than maximum change, the iteration stops and exit. If maximum change is 0.1 and a iteration wont lead to a change as big as 0.1, then iteration stops.

A WORKING CIRCULAR REFERENCE
To understand the circular reference more closely, let us construct a working circular reference. Go to cell A1 and enter **=B1** in cell A1. Then go to Cell B1 and enter **=A1+1.** Now a working circular reference is ready. Let maximum number of iteration be 1000 and maximum amount of change be 0.1

Each pressing of F9 key activates the counter and runs for 1000 times, advancing the value in B1 by 1000.

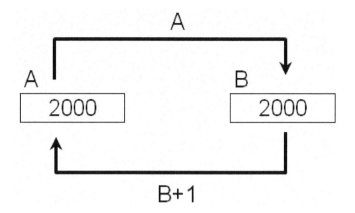

The cell A takes value from cell B but by adding 1 to what value is in B. At the same time cell B takes the current value of A. This constitute a cycle. This cycle repeat until maximum iteration or maximum change conditions set are reached.

Let maximum number of iteration be 32000 and maximum amount of change be 0.1. Each pressing of F9 key activates the counter and runs for 32000 times, advancing the value in B1 by 32000.

CASE 1
Maximum Iteration = 1
The counter advances one count for every activating of iteration.

CASE 2
Maximum Iteration = 1000, Maximum Change = 10
Maximum change is 10 and the change caused by the equation is only 1. The iteration does not proceed.

Form this example it is evident that the working of a circular reference depends up on maximum iteration and maximum change set and on actual increment triggered in each iterative step.

Blank Page

7   ADVANCED ITERATION

To use worksheet iteration for any useful purpose, we need a better control over iteration. In this session, we will see how advanced iteration can be performed and associated techniques.

APPLICATION OF ITERATION

We have already seen an example of iteration, which was a simple counter. This counter simply advanced its value and there is not much of a use of it more than to demonstrate a circular reference.

Most important use of iteration is to solve certain category of equations called **implicit equations.** As we will see in next chapters, the **primary condition** to solve an implicit equation by iteration within worksheet is that the equation should be re-arrangeable in the form x = g(x). The **second condition** is that the equation should converge to its solution in such a rearranged form!

Fortunately, this is not so strange as far as many practical applications are concerned. Let us see how excel can solve implicit equations by taking few examples.

SIMPLE ITERATION

Example 1

$$x / Cos\ (x) - 1 = 0$$

We get a 15 digit accurate solution for the above problem with these two simple steps.

STEP 1
On rearranging to get the style x = f(x),

$$x = Cos\ (x)$$

TEP 2
To check this equation will converge to a solution or not. The best and simple method to check this manually by a trial in an Excel worksheet in which

iteration is enabled (Maximum iteration be 1000 and maximum change be 0)

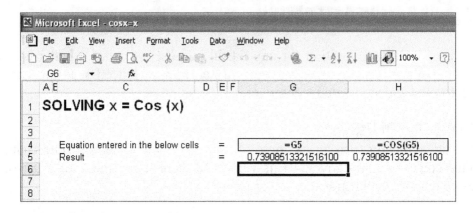

UNDERSTANDING THE CONVERGENCE

X-n PLOTS

A plot of attained value of x versus number of iterations (n) will give a clear idea on how equations converge to an accurate solution. This plot can be made by following procedure.

1) Set Maximum Iteration =1 (go to Tools>Options> Calculation)
2) Take down initial value of x , and note against n = 0
3) Press F9 once and note the result value of x against n=1
4) Repeat above and note the result against n=2, 3, ….
5) Plot x against n

Following is a x-n graph of the implicit equation x = Cos (x)

34

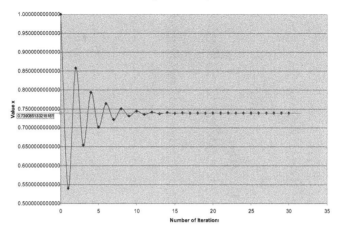

Convergence of x = cos (x

When an initial seed value of zero is considered, first iteration gives value 1 and subsequent iterations rapidly converge towards its solution. Thirty iterations resulted in 5 digit solution and ninety-three time iteration gave full 15 digit accurate solution.

VARIATION 1

It will be interesting to consider a slight variation of the equation.

$$x = 1.3 \cos (x)$$

Interestingly, this iteration took about 1000 iterations to get a 14 digit solution. The graph indicating initial points of convergence is presented below.

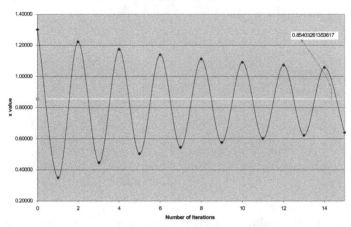

ANOTHER VARIATION OF THE SAME EQUATION

Now consider

$$x = 1.5 \text{ Cos } (x)$$

The plot of x value with number of iterations is as given below:

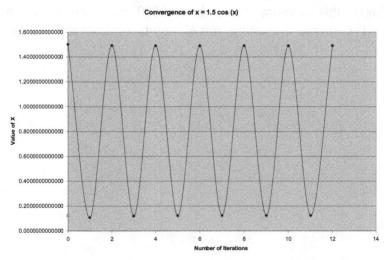

It is clear from the graph that this equation never converges to a constant

36

value for x.

From the above discussions, it is clear that depending up on the *nature* of equation, there can be a situation that *you not get a solution at all* or *you need to iterate thousands of times to reach an accurate solution, or it converges to satisfactory value in a few iterations.*

Larger the number of iterations gives more accurate the solution of implicit equation too. But iterating beyond certain number of cycles won't generally give any advantage.

MECHANISM OF ITERATIONS

The simple example above gave the result we needed; an accurate solution for an implicit equation. Two cells referred each other. *Left side cell* is X, and its initial value is zero. From this value, the dependent *right side cell* calculates Cos (x) which is Cos (0) = 1. Now the new x value is =1 after first iteration. During second iteration, the dependent cell calculates Cos function with this value of x, i.e. Cos (1) = 0.5403. This cycle is repeated till an accurate and stable x value arises. This x value is called "the solution" of the implicit equation.

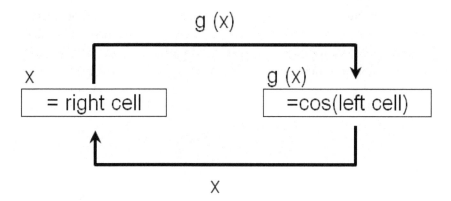

From this figure, we can note following points.
1) We have left cell, which is x, and gets x values from right cell
2) The value of x is zero initially
3) We have right cell which is a function of x, g(x). In this case, cos(x). This function operate on value of x from left cell
4) This cycle repeat and stabilise giving a x value called "solution" of the implicit equation.

5) The solution we get in seconds is x = 0.73908513321516100
6) It takes about 100 iteration to get this solution and complete instantly
7) Note that Tools>Calculations>iterations setting are as below:
   Maximum number of iteration=100, Maximum change = 0

TIME CONSUMED FOR ITERATION

Iteration consumes more time than normal worksheet calculations, which depends up on your processor speed, memory installed and other programs running in the background. In a typical experiment, a simple iteration of 32000 times took only 4.90 seconds. Such 10 calculations in same worksheet did not result in 10 fold increase in time to finish, instead finished in 15.52 seconds.

OPTIMUM NUMBER OF ITERATIVE CYCLES

As iterations are time consuming (though only in terms of seconds), it is always better to confine the iterative cycles to a balance between precision needed and time consumed. Too much of precision in the order of 15 significant digits may not be needed in scientific or engineering calculations.

Each implicit equation needs a test for its convergence pattern to decide a moderate balance between precision and number of iterations. For a pipe *friction factor* calculation, maximum precision required will be 5 digits and so number of iterations need not be in several thousands. The test can be performed by selecting Maximum Iteration =10 and pressing F9 key several times until you reach necessary stability. If 15 pressing of F9 key resulted in a 10 digit precision (evident by stabilizing of left 10 digits), an average of 150 iteration (10 cycles per pressing of F9 and 15 such pressing means 150 cycles) is sufficient for the calculation.

8   CONTROLLING ADVANCED ITERATIONS

In last chapters, we saw how iteration of a simple equation x = Cos (x) progress. Only way by which we could control its progress was by adjusting Maximum Iteration, Maximum Change and by pressing F9 key for performing each batch of iterations. There should be a better way to control and closely examine, and understand iterations. This will not only help to customize the iteration, that by improving the efficiency along with several other advantages in testing and adapting iterative strategy based on characteristics of the implicit equation.

CONCEPTS
Following are the basic facilities and concepts needed to make a iteration user-friendly enabling proper control over each iteration equation in the worksheet.

1) ITERATING CELLS
For our iterating technique, we need a cell which we call x and another cell which we call f(x). Example, in the implicit equation x = Cos (x). These cells interdependent and having *circular reference* to each other is called **Iterating cells.**

2) SEED
Seed is the initial value for x that will be used to start iteration. Seed is important that correctly providing a seed will only give a correct solution for an implicit equation. Ideally the seed should be an "approximate" value assumed for the solution.

For the solution of x = Cos (x), the seed can be the default value of any cell in excel which is zero. But in specific implicit equations, it is important that which initial value is assumed, for example –ve value, -1, 0, +1, +value, negative and positive fractions of pi etc. A correct seed only can give rise to a correct divergence. Most simple method to test this is by trial and error.

3) COUNTER
A counter should be there near a iteration to see how many times iteration was actually performed. This helps us to understand how value of x proceeds with number of iterations.

4) CONTROLS

Control is a cell that may take value of 1, 2 or 3 based on your ON, OFF or RESET options selected in FORM buttons. How to proceed with iteration should be specified using IF statements in the iterating cells.

ON and OFF CONTROL

In a user-friendly worksheet, we should be able to trigger and arrest iteration by input selections. Once iteration is arrested, any further change in the worksheet should not trigger iteration.

RESET CONTROL

Reset button is a useful way to bring back the iteration to its initial state and the initial x value equal to the **Seed**.

From this state a different seed can be input and switch on the iteration. Besides this, any other parameters like ITERATION CYCLES can be changed after a RESET

5) ITERATION CYCLES

We have seen that Tools>Options>Calculation gives dialogue box to specify Maximum Iterations and Maximum Change. It is not convenient to specifically control a particular iteration by this method. For this purpose, we need to set up a control box in which we can specify specific numbers of Iteration Cycles.

6) INDICATOR

An indicator may be needed which indicate a green signal when required precision is reached.

CREATING A SEED

As already explained, seed is the starting value to be assumed for x. In a typical worksheet, we have shown iteration between cells E5 and F5. In E5 there is a reference to F5 (i.e. equation entered in E5 is = **F5**). In F5, the equation is =**Cos (E5)** thus triggering the iteration.

Seed can be introduced by using an IF statement in E5 which should act based on the **Control**. Let us see how. Let **Control** cell be H5 and **Seed** be I5.

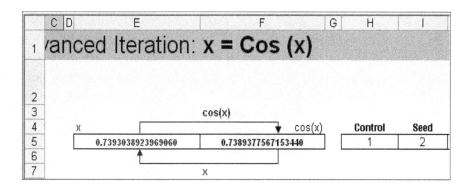

Only thing we have to do is to introduce a IF condition in the x cell that if control value is 3, use seed value as x, otherwise use value from f(x) cell itself.

IF(**control=3**, use seed, use the new Cos(x) value)

In the cell E5, the required equation is

=IF(H5=3,I5,F5)

This equation ensures that you can introduce a new SEED value and activate its introduction to iteration by entering 3 in control. There is a better method to enter 3 in the cell by a mouse-click which we will discuss later.

INTRODUCING *ON* AND *OFF* CONTROL

You might have observed that any change in any cell in the worksheet could activate a batch of iteration even after sufficient precision is reached. This way many iterations initiate each time you alter or enter a value in any cell, causing some delay in recalculation and some instability feel seeing changing values of solutions (which are actually progressing to a more accurate value, but mostly precision beyond our requirement)

An OFF and ON switch can be set up in the worksheet to halt/allow continuation of iteration by introducing an IF condition in the *f(x)* cell telling that the cell need to return *f(x)* only if **Control** is 1 and it should return value of x from the left cell if control is made 2.

The IF statement in the cell F5 will be as below
$$IF(H5=1,Cos(E5),E5)$$

41

This IF statement looks for value of **Control** and decided what value to bring. If Control is 1, it brings x value from E5 and converts it to Cos (x). Otherwise it simply returns E5, i.e. current x value itself without operating the Cos function and hence the iteration dies. The effect of changing Control value from 1 to 2 is shown in the following figure:

When Control = 1 (the iteration will continue)

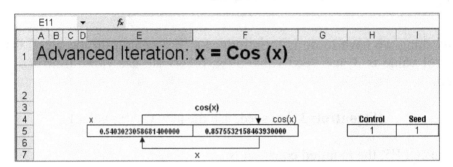

When Control = 2 (the iteration is halt and further change of value cant occur)

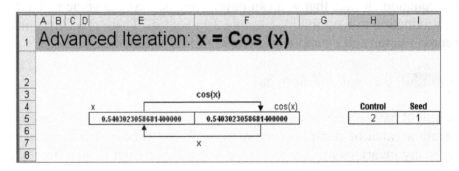

CREATING A COUNTER

A counter should count each time iterative cycle is executed and stop if the iteration is stopped and the **Control** is 2. The counter can be set up in cell J5 with following equation with an IF function in it

We need excel to perform a logic: If (Control=2, no need to count further, otherwise count one for each iteration happening in worksheet)

In the Cell J5,

$$=IF (H5=2,J5 ,J5+1)$$

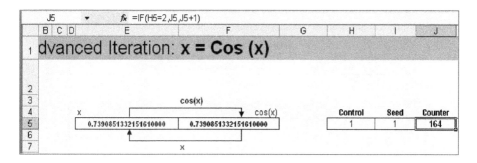

CREATING A INDICATOR

As we already discussed, iterating cells are the pair consisting the cell x and the cell *f(x)*. The indicator should indicate green when **expected equality** is reached between (x) and new value of *f(x)*

Indicator for 15-Digit equality

We have already seen that maximum precision of numbers in excel is 15. So a check of equality of iterating cells can be used to activate indicator.

Let the cell G5 be the indicator cell. First we calculate difference between iterating cells. Following equation will return the absolute value of the difference between iterating cells. If this results zero, means iteration has resulted a 15 digit precise result.

=ABS (E5)-ABS (F5)

Now we can set up indication by the facility **Format > Conditional Formatting.** First click the cell G5 where we need the indicator. Then go to Format > Conditional Formatting. Select equal to and enter 0 as the condition as below. Go to Format button and adjust cell shade needed. (Selected green in the example)

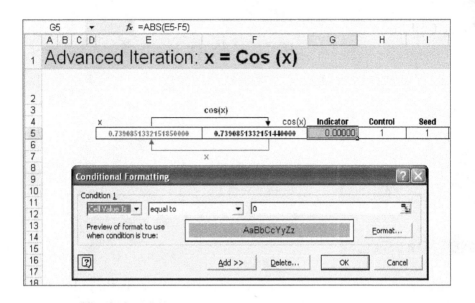

The indicator will display green once iterating cells equals.

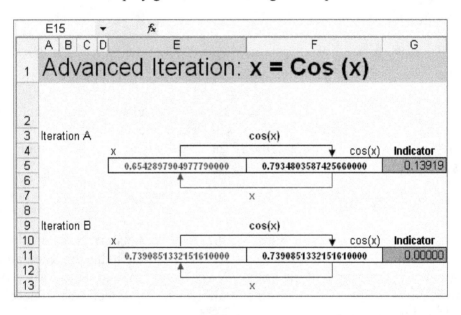

Iteration A is incomplete. So it is orange color, the basic format selected for that cell. But when the cell value becomes zero after complete iterations, the cell indicates zero, informing us quickly that the full precision is achieved.

Indicator for 5 digit precision:

In conditional format, select **less than 0.000001** instead of **equal to 0**. Then the cell turns green when cell value drops less than **0.000001,** which means 5 digits of equality between the iterating cells.

BUTTON SELECTOR FOR CONTROL CELL

As already discussed, the iteration is controlled using a **Control cell** which need user input 1, 2 or 3 for ON, OFF or RESET. Remembering and entering is not very easy. A box consisting of 3 buttons from which you can manually select ON/OFF/RESET option will be a great convenience.

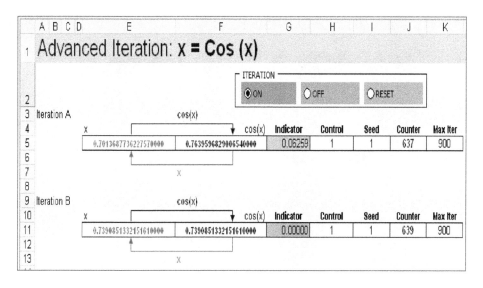

Here is the procedure to create this button box.

1   Set Maximum Iteration to 100 and Maximum Change to 0. Maximum change to zero means iteration will run even if there is no change of value after each cycle of calculation.

2   Set up iteration using E5 as x and G5 as Cos (x) as discussed in the previous chapter.

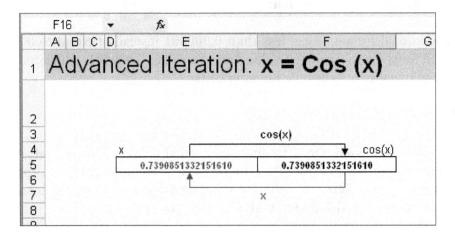

3   Mark the cell H5 as CONTROL.

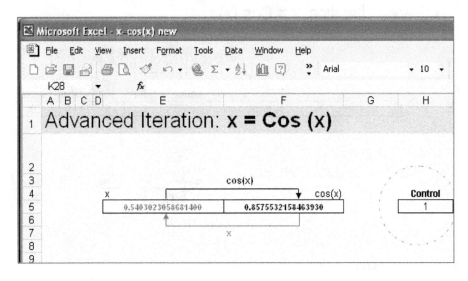

4   Go to View > Tool bars and select Forms. You will see the following toolbar. Then click the **Group Box** indicated below.

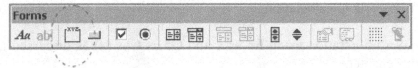

5   Go to worksheet and draw a rectangle **Group Box** as below.

6   Insert Option Buttons in to this rectangle by selecting it from Forms toolbar (marked blue)

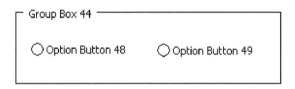

7   The Group Box will look like:

Group Box 44

○ Option Button 48   ○ Option Button 49

8   Click on any of the button and it will now look as below

Group Box 44

◉ Option Button 48   ○ Option Button 49

or

Group Box 44

○ Option Button 48   ◉ Option Button 49

9   Now select any of the button rectangle as below

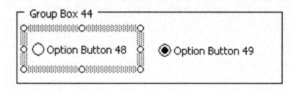

10   Link the button rectangle to see the menu as below

11   Click Format Control

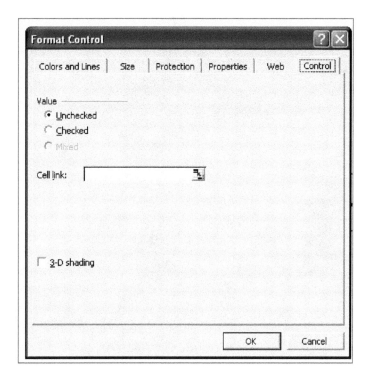

12   Click Cell link and enter H5 in the box. Click OK button at bottom of the Form Control dialogue box

13   Rename first button as ON, and second button as OFF

14   Include one more button and name it RESET

Your buttons can be edited and formatted like this, but in different colors:

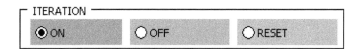

Blank Page

9   SOLVING COLEBROOK EQUATION

With the background we laid using the simple case of x = cos(x), now we come to the main topic of this book.

DARCY-WEISBACH EQUATION

In 1957, Henry Darcy came up with a similar but some what different equation "for old rough pipes" under turbulent flow conditions. Also he proposed a factor which depends on pipe roughness and diameter. Some references says that Darcy never proposed factor f as proposed by Weisbach but the friction factor f is known after his name now!. Julius Weisbach (1806-1871), a native of Saxony proposed in 1845 the equation for h_{loss} as we know now as "Darcy-Weisbach equation":

$$h_{loss} = f \ (L/D) \ u^2/2g$$

Where f = friction factor or friction coefficient.

The term $u^2/2g$ is known as "Velocity Head" where
u = flow velocity and
g = acceleration of gravity.
L= length of duct/pipe
D is hydraulic diameter of the duct.

MOODY DIAGRAM

The condition for turbulent flow is when Reynolds Number is above that of transition range. For almost all of industrial duct systems and industrial fluid circulations and transportation, the velocity need be sufficiently high that common condition that exist is turbulent flow condition, i.e. Re > 4000.

Lewis F. Moody (Professor, hydraulic engineer, Princeton University) in ASME Transactions of November 1944 published Moody's work (and Moody Diagram), in his article entitled "Friction Factors for Pipe Flow". It is a plot of Fanning friction factor against Reynolds number. *It is a graphical solution for Colebrook equation.*

This has become basis for many of the calculations on friction loss in pipes, duct work and flues. Moody references the work of C. F. Colebrook and C.

M. White in developing Moody Diagram. Crane TP410 or any hydraulics textbook gives Moody Diagram. It is also available in internet for free.

COLEBROOK EQUATION
There are many formulas for calculating friction factor for turbulent conditions. Colebrook (1938) developed the best-known formula presently known after his name.

$$\frac{1}{\sqrt{f}} = -2 \; \log\left(\frac{e}{3.7} + \frac{2.51}{Re \; \sqrt{f}} \right)$$

Where e is relative roughness
f is the friction factor
Re is re Reynolds numbers.

It is a hydraulic criterion, which specifies the laminar or turbulent characteristics regime of the airflow. The Reynolds number, Re is defined as,

$$Re = \frac{\rho V D}{\mu}$$

ρ : Fluid density
μ : Fluid absolute viscosity
V: The average flow velocity
D: Pipe diameter

Colebrook equation is a "understood though not directly expressed" equation otherwise known as implicit equation. Solving such equation requires that you assume a value for f to start the iteration. The assumed values converge in few (less than 100) iterations.

Iterations are not easy unless you know sequential programming in FORTRAN or VBA. Many attempts to solve Colebrook equation by engineers are not perfect, as they need either a deep knowledge of Visual Basic or have to use tools like Goal Seek available in Excel, which makes multiple solving within a single sheet highly time consuming and non-automatic (requiring manual interventions).

We researched for a simple technique to solve implicit equations within the worksheet and this book is a result of this. We extended the technique to various scientific/engineering systems like **van der Waal** equations successfully.

RELATIVE ROUGHNESS

The term ε in the above term is relative roughness which is found by experiments (fitting the pressure-drop data in to Colebrook equation) or in comparison to the roughness parameters of pipes made of various materials reported in literature. Also it is to be remembered that relative roughness is a parameter of importance only for turbulent flow and this is the condition in all industrial exhaust ducts and mostly in piping

Relative Roughness suitable for substituting in to Colebrook equation for commercial pipes is presented in various literature. Following table gives the typical values that can be used to solve Colebrook equation. It is emphasized here that no major error will be introduced by a slight wrong judgment. However, for accurate and reliable values, consult literature and research publications or tube or manufacturers recommendations.

SOLVING COLEBROOK EQUATION

Steps described in solving Colebrook equation is similar to what we adopted in the case of the solving the equation $x = \text{Cos}(x)$.

STEP 1: Rewrite Colebrook equation in the form of x = function of x

STEP 2: Iterate these equations in Excel

Colebrook equations converge very well to solutions quickly, in about 30 iterations it reaches 15-digit precision! Please see next chapter for details

Blank Page

10   FIXED POINT ITERATIONS – The Basics

INTRODUCTION
This chapter gives you mathematical background required to understand iterative solutions discussed in this book. Those who have no academic interest can skip this chapter.

A **nonlinear equation** is where the variable(s) to be solved for, cannot be written as a linear combination of independent components. A nonlinear equation is an equation which won't form a straight line when inputs and outputs are plotted. Some of the physical systems in this world are nonlinear. In such cases, a change in one parameter will not result a linear change in other parameter. **Transcendental equations** are nonlinear equations which may contain trigonometric, logarithmic or exponential functions in it.

Implicit equations are a special case of nonlinear equations which contain nonlinear relation between parameter x and the output function of that parameter, f(x). Colebrook-White equation is a typical example.

OTHER EXAMPLES FOR IMPLICIT NONLINEAR FUNCTIONS

$$x = 0.9 \; Cos \; (x) \; or \; f(x) = x - 0.9 \; Cos \; (x)$$

$$x = 1 - [(0.5)^x - 0.541]/0.3$$

$$x = x^3 - 1$$

$$x = e^{-0.7x}$$

SOLUTION

A linear equation can be solved simply by substitution method. But the solution of nonlinear systems of algebraic equations is frequently not possible using formal analytic methods, as it is in for linear systems. Thus it is necessary to resort to *numerical methods* to solve such systems

Many problems in engineering and science require a **solution** for nonlinear equations. Consider a function of x, f(x) = 0. A solution for this function is a

number which used as a value for x satisfies the equation x = F(x). Finding solution (a value of x) for which x = F(x) is equivalent to finding root of the equation f(x) = 0

Colebrook equation is a function of friction factor. To solve Colebrook equation for its 'x' (friction factor) is to rearrange the Colebrook equation in the form of x = F(x) and to find the value of 'x', the friction factor which satisfies this.

ITERATION

As the name suggests, iteration means the calculation process is repeated until an answer is achieved. Iterative techniques are used to find roots of equations, solutions of linear and nonlinear systems of equations, and solutions of differential equations.

The iterative scheme to find solution by fixed-point iteration is as follows
- Assume a numerical value for x
- Calculate g(x)
- Use this g(x) as new value of x
- Calculate g(x) from this new value..
- Repeat this...

FIXED POINT
Fixed point of a function f(x) =0, which is rearranged to the form x = g(x) is a **number** p such that p = g (p). It is a solution of the equation x = g(x). The fixed point of x = g(x) is also a zero (root) of the corresponding equation f(x) = 0.

Example:

Function: f(x) = 0,	x – cos (x) = 0
Rearranged: x = g(x)	x = cos (x)
Fixed point = p, so that p = g(p)	0.739303892396906
p is solution of p = g (p) and root of f(x) =0	0.739303892396906 is solution of **x = cos (x)** and root of **x – cos (x) = 0**

GRAPHICAL SOLUTION

Graphically, a solution can be obtained by intersection of curves $Y = x$ and $Y = g(x)$ plotted against x. The intersection is at $x = g(x)$ which is the solution of the equation.

COLEBROOK EQUATION AS f(x)=0

$$\frac{1}{\sqrt{f}} = -2 \, \log\left(\frac{e}{3.7} + \frac{2.51}{Re \, \sqrt{f}}\right)$$

Rearranging,

$$1/f^{0.5} - 2 \log [e/3.7 + 2.51 / (Re \; f^{0.5})] = 0$$

We need to rearrange the equation to the form $x = g(x)$

$$f = (-1/\{2 \log [e/3.7 + 2.51 / (Re \; f^{0.5})]\})^2$$

X = g(X) PLOT

For e = 0.01 and Re = 500, following plot can be constructed which intersect at an approximate solution which is around 0.0472 visually from graph. Note that the solution is not very accurate and this is a tedious method. A single solution need so much of calculations. More over, the solution can not be considered for a further calculation, unless the approximation is manually fed.

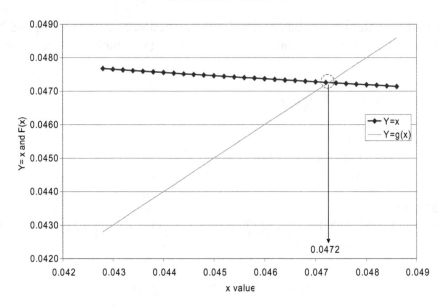

Plots of x versus Y=x and Y=F

CONVERGENCE CONDITIONS

An implicit equation or a nonlinear function f(x) = 0 can be rewritten to x = g(x) form in many ways.

For example, take the case of

$$f(x) = x^3 + x + 1 = 0$$

Rewriting this to x = g(x) form has following forms of possibilities.

$$x = x^3 - 1$$
$$x = 1/(x^2 - 1)$$
$$x = (1 + x)^{1/3}$$

All these equations wont converge. How to know which equation among this will diverge to a solution if iterated?

For a function F(x) = 0 to |F'(a)| should be < 1 where a is a approximation for the solution and F' is the first derivative of F(x)

This condition is satisfied by only the third one above which is $x = (1 + x)^{1/3.}$ To elaborate on this is beyond the scope of this book.

58

For a person with no expertise and experience in mathematics to understand and apply the above condition, this is possible only by a trial test by constructing iteration in worksheet and seeing how the value proceeds.

TESTING THE PROGRESS OF ITERATION

To know the equation converge or not, a initial test is only the simplest way. For this, Tools>Options>Calculation>Iteration>Maximum Iteration be set to 1. Press F9 each time and note the vale in the cell containing x. If it converge the value proceeds towards a stable solution. As already pointed out in a previous chapter, slight change in a equation makes it con-converging (x = cos(x) converge, but x = 1.5 cos(x) not!)

11   COLEBROOK EQUATION - SOME IMPORTANT POINTS

Turbulent fluid flows in pipes and open channels play an important role in hydraulics, chemical engineering, transportation of hydrocarbon, air duct sesign, etc.

1

These flows induce a significant loss of energy depending on the flow regime and the friction on the rigid boundaries. When fluid flows through a pipe, friction between the pipe wall and the fluid works against the flow and is one of the most important parameters in determining a pipeline's capacity

2

The head loss (h_f) due to friction undergone by a fluid motion in a pipe is usually calculated through the Darcy-Weisbach relation.

$$h_{loss} = f \ (L/D) \ u^2/2g$$

Where f = friction factor or friction coefficient.

The term $u^2/2g$ is known as "Velocity Head" where
u = flow velocity and
g = acceleration of gravity.
L= length of duct/pipe
D is hydraulic diameter of the duct.

3

The friction factor (f) is a measure of the shear stress (or shear force per unit area) that the turbulent flow exerts on the wall of a pipe and expressed in dimensionless form as $f = \tau/\rho \bar{u}^2$, where, τ is the shear stress, ρ is the density of the liquid that flows in the pipe and $\bar{u}$ the mean velocity of the flow.

4

For laminar flow (Reynolds number, $R \leq 2100$), the friction factor is linearly dependent on R, and calculated from the well-known Hagen-Poiseuille equation:

$$\lambda = 64\ /R$$

Where, R, the Reynolds number, is defined as $\bar{u}\ D/\ v$ where v is the kinematic viscosity $= \mu/\rho$, the ratio of viscosity and density.

Flow in offshore gas pipelines is characterized by high Reynolds numbers, typically highest, in the order of 10^7, due to the low viscosity and the relative high density of natural gas at typical operating pressures (100-180 bar). For normal liquid lines, the Reynolds number is normally in the range of 5×10^4 to 1×10^6

5

In turbulent flow ($R \geq 4000$), the friction factor, f depends upon the Reynolds number (R) and on the relative roughness of the pipe, e/D, where, e is the average roughness height of the pipe

6

When k is very small compared to the pipe diameter D i.e. $e/D \rightarrow 0$, and f depends only on R.

7

SMOOTH LAMINAR FLOW REGIME: When k/D is of a significant value, at low R, the flow can be considered as in smooth regime (there is no effect of roughness). In a smooth pipe flow, the viscous sub layer completely submerges the effect of e
on the flow. In this case, the friction factor f is a function of R and is independent of the effect of e on the flow.

8

CRITICAL ZONE: Fluids with a Reynolds number between 2000 and 4000 are considered unstable and can exhibit either laminar or turbulent behavior. This region is commonly referred to as the critical zone and the friction factor can be difficult to accurately predict. Judgment should be used if accurate predictions of fluid loss are required in this region. Colebrook equation can be examined to get a rough estimate of friction in this region.

9

If the Reynolds number is beyond 4000, the fluid is considered turbulent and the friction factor is dependent on the Reynolds number and relative roughness

TRANSITION REGIME: As R increases, the flow becomes transitionally rough, called as transition regime in which the friction factor rises above the smooth value and is a function of both k and R.

ROUGH TURBULENT REGIME: As R increases more and more, the flow eventually reaches a fully rough regime in which f is independent of R.

In this zones the friction factor f is calculated by Colebrook's equation (Colebrook 1938-39)

10

Nikuradse (1933) verified the Prandtl's mixing length theory and proposed the following universal resistance equation for fully developed turbulent flow in smooth pipe;

$$1/ \sqrt{f} = 2 \log (R \quad \sqrt{f}) - 0.8$$

11

In case of rough pipe flow, the viscous sub layer thickness is very small when compared to roughness height and thus the flow is dominated by the roughness of the pipe wall and f is the function only of k/D and is independent of R. The following form of the equation is first derived by Von Karman (Schlichting, 1979) and later supported by Nikuradse's experiments

$$1/ \sqrt{f} = 2 \log (D/e) + 1.74$$

12

COLEBROOK EQUATION: The Colebrook–White equation estimates the (dimensionless) Darcy–Weisbach friction factor f for fluid flows in filled pipes. For transition regime of flow, in which the friction factor varies with both R and e/D, the equation universally adopted is due to Colebrook and White (1937) proposed the following equation

$$\frac{1}{\sqrt{f}} = -2 \, \log\left(\frac{e}{3.7} + \frac{2.51}{Re \, \sqrt{f}} \right)$$

e is relative roughness
f is the friction factor
Re is Reynolds numbers.

The Colebrook equation can be used to calculate the friction coefficients in different kinds of fluid flows - air ventilation ducts, pipes and tubes with water or oil, compressed air and much more.

13

Colebrook Equation covers not only the transition region but also the fully developed smooth and rough pipes. By putting $e \rightarrow 0$, this reduces to equation for **smooth pipes** and as R→∞, it forms equation for **rough pipes**.

Colebrook's transition curve merges asymptotically into the curves representing laminar and completely turbulent flow.

14

In a smooth pipe flow, the viscous sub layer completely submerges the effect of k on the flow. In this case, the friction factor f is a function of R and is independent of the effect of k on the flow.

15

In case of rough pipe flow, the viscous sub layer thickness is very small when compared to roughness height and thus the flow is dominated by the roughness of the pipe wall and f is the function only of k/D and is independent of R.

16

The Colebrook equation has two terms. The first term, (e/D)/3.7, is dominant for gas flow where the Re is high. The second term, 2.51/Re $\sqrt{f}$, is dominant for fluid flow where the relative roughness lines converge (smooth pipes).

17

Moody (1944) presented a friction diagram for commercial pipe friction

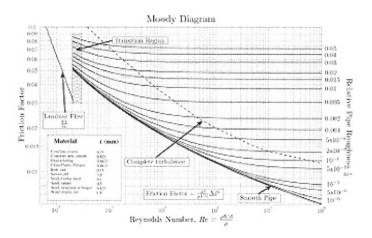

factors based on the Colebrook–White equation, which has been extensively used for practical applications. This is known as Moody's Chart or Diagram.

In the "Complete Turbulence" region, these lines are "flat", meaning that they are independent of the Reynolds Number. In the "transition Zone", the lines are dependent on Re and . When the lines converge in the "smooth zone" the fluid is independent of relative roughness.

Recently it has come to know that Moody diagram is only +/-15% accurate. Glenn O. Brown (professor, Oklahoma State University, OK) recently in his published on internet wondered why, for more than 6 decades, the Moody diagram was used unmodified and still in use same way.

18

The traditional Moody diagram is still a much widely used method to determine f, even though some major limitations apply to it: It over-estimates the 'smooth pipe'-friction for Re > 10^6, and it is very inaccurate around Re = 10^4

19

Because of Moody's work and the demonstrated applicability of Colebrook-White equation over a wide range of Reynolds numbers and relative roughness value *e/D*, COLEBROOK equation has become the **universally accepted standard** for calculating the friction factors for rough pipes.

20

Since the mid-1970s, many alternative explicit equations have been developed to avoid the iterative process that is inherent in Colebrook- White equation. These equations give a reasonable approximation; however, they tend to be less universally accepted.

21

One of the major draw back of Colebrook equation is that it is implicit! However, implicit equations can be solved in Excel, which has been discussed in this book in great detail.

22

Some researchers have found that the Colebrook–White equation is inadequate for pipes smaller than 2.5 mm

Zagarola (1996) has indicated that the Prandtl's law of flow in smooth pipes was not accurate for high Reynolds numbers and the Colebrook-White correlation (which was based on the Prandtl's law of flow) is not accurate at high Reynolds numbers[4].

22

The major cost of piping and pumping system is always thought to be the capital investment on the piping and pumping units. However, the recurrence expenditure on the power consumption has to be given due consideration because generally the recurrence expenditure over the years far outweigh the

---

[4]Zagarola, M. V., "Mean-flow Scaling of Turbulent Pipe Flow," Ph.D.thesis, Princeton University, USA, 1996.

capital investment. Hence the major length of the pipeline is designed with a view to minimize both the power requirement and pipe material simultaneously[5].

23

Idelchik summarizes roughness factors for 80 materials including metal tubes, conduits made from concrete and cement; and wood, plywood, and glass tubes.

Idelchik, I.E., M.O. Steinberg, G.R. Malyavskaya, and O.G. Martynenko. 1994. Handbook of hydraulic resistance, 3rd ed. CRC Press

24

Typical values of absolute roughness are 0.0015 mm for PVC, drawn tubing, glass and 0.045 mm for commercial steel/welded steel and wrought iron

25

Fast, accurate and robust resolution of the Colebrook equation is necessary to design computations and flow simulations. A fast and simple computational method to find out solution for Colebrook as presented in this book, is expected to effectively contribute to such optimizations.

26

Notwithstanding certain shortcomings as reported by certain researchers, the Colebrook equation is currently the most accepted equation for calculating the friction factors in most of the industrial flow calculations.

---

[5]**"Energy Efficient Pipe Sizing and Piping Optimization"**– A book by the same author. Visit www.ColebrookEquation.com for more details

EXCEL WORKSHEETS

PURCHASE OF WORKSHEETS
In case you wish to purchase a copy of the worksheet, which you can use with any converging implicit equation, the same can be had from our publication division.

1000 Colebrook solve instances in a Excel Worksheet : 20 US$

Please mail to frpcad@gmail.com for your copy. Please indicate the identification number of purchase of this book.

MAKING YOUR OWN WORKSHEET
Specifically for your own calculation, you can make any number of worksheets in your own style and flexibility using the knowledge laid out in this book.

PROGRAMMING ASSISTANCE
In case you prefer us to make the worksheet for you, we provide Excel Programming Assistance.
Contact frpcad@gmail.com for specific offers.

WE REQUEST YOUR SUGGESTIONS

We request your suggestions to improve this book and we will definitely assist you personally in solving implicit equation in your worksheet.
Please feel free to contact the author!
-Author

OTHER BOOKS OF INTEREST

"Universal Unit Conversion Capability Within Excel Worksheet"
by M Anil Kumar

A simple step by step guide about Units, Dimensions, Unit Conversion principles and procedure to make your worksheet calculation smart by setting up universal unit converter capabilities. No VBA/Macros/UDF/Add-ins!

"Energy Efficient Pipe Sizing and Piping Optimization"
by M Anil Kumar

Piping systems are often sized by guesstimates. Importance is solely given for capital expenditure and not on the life-cycle power consumption costs. In this word, when energy becomes scarcer and costly, it is high time to think of energy huge saving and return of investment thus possible.

Visit www.ColebrookEquation.com for more details

"Duct Design Fundamentals" For Industrial Engineers and Scientists.

Principles of exhaust ducting design and simple balancing procedures using Excel worksheets.

SOLVING COLEBROOK EQUATION

for

Pipe Design, Sizing and Optimisation Calculations

M Anil Kumar

(MS Chemical, IIT Madras, Industrial Consultant)

A.N. NIVAS - KADIRUR
INDIA 670642
++91 9020683510
www.ColebrookEquation.com